Percentages

By Robert Watchman

An Easy Steps Math series book

Books in the Easy Steps Math series

Fractions
Decimals
Percentages
Ratios
Negative Numbers
Algebra
Master Collection 1 – Fractions, Decimals and Percentages
Master Collection 2 – Fractions, Decimals and Ratios
Master Collection 3 – Fractions, Percentages and Ratios
Master Collection 4 – Decimals, Percentages and Ratios

More to Follow

Contents

Introduction

This series of books has been written for the purpose of simplifying mathematical concepts that many students (and parents) find difficult. The explanations in many textbooks and on the Internet are often confusing and bogged down with terminology. This book has been written in a step-by-step 'verbal' style, meaning, the instructions are what would be said to students in class to explain the concepts in an easy to understand way.

Students are taught how to do their work in class, but when they get home, many do not necessarily recall how to answer the questions they learned about earlier that day. All they see are numbers in their books with no easy-to-follow explanation of what to do. This is a very common problem, especially when new concepts are being taught.

For over twenty years I have been writing math notes on the board for students to copy into a note book (separate from their work book), so when they go home they will still know how the questions are supposed to be answered. The excuse of not understanding or forgetting how to do the work is becoming a thing of the past. Many students have commented that when they read over these notes, either for completing homework or studying for a test or exam, they hear my voice going through the explanations again.

Once students start seeing success, they start to enjoy math rather than dread it. Students have found much success in using the notes from class to aid them in their study. In fact students from other classes have been seen using photocopies of the notes given in my classes. In one instance a parent found my math notes so easy to follow that he copied them to use in teaching his students in his school.

You will find this step-by-step method of learning easier to follow than traditional styles of explanation. With questions included throughout, you will gain practice along with a newfound understanding of how to complete your calculations. Answers are included at the end.

Chapter 1

Percentage Basics

The word *percent* is derived from the Latin *per centum*, which means 'by the hundred'. Today we use *percent* to mean '*per hundred*' or '*out of one hundred*' or '*over one hundred*'. The symbol used to abbreviate percent is (%) and is written directly after the number it describes.

Once you know your fractions and decimals, percentages become easy to calculate and convert. This is because the conversion process uses multiplication and division of fractions and decimals (see the Easy Steps Math Fractions and Easy Steps Math Decimals books).

A percentage is basically any value out of one hundred.

Twenty-five percent (25%) means 25 out of 100. This can be written as the fraction $\frac{25}{100}$, or as the decimal 0.25.

Seventy-eight percent (78%) means 78 out of 100. This can be written as the fraction $\frac{78}{100}$, or as the decimal 0.78.

One hundred and twenty percent (120%) means 120 out of 100. This can be written as the fraction $\frac{120}{100}$, which is simplified down to $1\frac{1}{5}$, or as the decimal 1.2, and so on.

There are a few percentages that you should know off the top of your head. These come in very handy in everyday life because businesses use these regularly for discounts or interest rates etc.

Percentage	5	10	20	25	$33\frac{1}{3}$	50
Fraction	$\frac{1}{20}$	$\frac{1}{10}$	$\frac{1}{5}$	$\frac{1}{4}$	$\frac{1}{3}$	$\frac{1}{2}$
Decimal	0.05	0.1	0.2	0.25	0.333	0.5

Percentage	$66\frac{2}{3}$	75	90	100	150	200
Fraction	$\frac{2}{3}$	$\frac{3}{4}$	$\frac{90}{1}$	$\frac{1}{1}$	$\frac{150}{1}$	$\frac{2}{1}$
Decimal	0.666	0.75	0.9	1.0	1.5	2.0

Chapter 2

Percentages and Fractions

Firstly it is important to know how to **convert a percentage to a fraction**. This is one of the easiest conversions because all you do is put the percentage over 100 and drop the percent sign then simplify.

Examples,

a) Convert 45% to a fraction.

Step 1. Put the percentage over 100 and drop the % sign.

$$\frac{45}{100}$$

Step 2. Simplify

$$\frac{45^{\div 5}}{100_{\div 5}} = \frac{9}{20}$$

So 45% is the same as $\frac{9}{20}$

b) Convert 137% to a fraction

Step 1. Put the percentage over 100 and drop the % sign.

$$\frac{137}{100}$$

Step 2. Simplify

$1\frac{37}{100}$ (See the rules for simplifying improper fractions in the Easy Steps Math Fractions book).

Now you try these questions.

Convert these percentages to fractions

a) 11%

b) 67%

c) 23%

d) 36%

e) 170%

f) 129%

g) 315%

h) 240%

i) 360%

j) 400%

Occasionally it becomes necessary to **convert a fraction to a percentage**. Since this method uses multiplication of fractions, it is also a simple process. All you do is multiply by 100 and put the percentage sign $(\%)$ after the answer.

Examples

a) Convert $\frac{9}{20}$ to a percentage.

Step 1. Write out the fraction and multiply by 100

$$\frac{9}{20}\times100$$

Step 2. Change the 100 to a fraction. (Remember from the Easy Steps Math Fractions book that a whole number can be written as a fraction when you put it over 1).

$$\frac{9}{20}\times\frac{100}{1}$$

Step 3. Cross simplify the two fractions and multiply the numerators then the denominators.

$$= \frac{9}{_1\cancel{20}} \times \frac{\cancel{100}^5}{1} = \frac{9}{1} \times \frac{5}{1}$$

$$= \frac{45}{1}$$

Step 4. Simplify your answer and add the % sign

$$= 45\%$$

Therefore $\frac{9}{20} = 45\%$

b) Convert $\frac{3}{5}$ to a percentage.

Step 1. Write out the fraction and multiply by 100

$$\frac{3}{5} \times 100$$

Step 2. Change the 100 to a fraction.

$$\frac{3}{5} \times \frac{100}{1}$$

Step 3. Cross simplify the two fractions and multiply the numerators then the denominators.

$$= \frac{3}{\cancel{5}_1} \times \frac{\cancel{100}^{20}}{1} = \frac{3}{1} \times \frac{20}{1}$$

$$= \frac{60}{1}$$

Step 4. Simplify your answer and add the % sign

$$= 60\%$$

Therefore $\frac{3}{5} = 60\%$

Convert these fractions to percentages.

a) $\frac{9}{10}$

b) $\frac{7}{20}$

c) $\frac{1}{2}$

d) $\frac{12}{25}$

e) $\frac{9}{30}$

f) $\frac{28}{40}$

g) $\frac{7}{4}$

h) $\frac{3}{50}$

i) $\frac{4}{5}$

j) $1\frac{1}{4}$

A **fractional percentage** is a percentage that includes fractions, such as $\frac{1}{4}\%$. To convert this to a fraction requires an extra step or two. Once you learn this, you can do it in your head.

Remember from the Easy Steps Math Fractions book, that:

a) a fraction is a division, and

b) a percentage is a fraction out of 100.

Putting these two bits of information together we can work out that $\frac{1}{4}\%$, is the same as $\frac{\frac{1}{4}}{100}$ (one quarter over one hundred). Using division of fractions we can work out that this is the same as

$$\frac{1}{4} \div 100$$

$$= \frac{1}{4} \div \frac{100}{1}$$

$$= \frac{1}{4} \times \frac{1}{100}$$

$$= \frac{1}{400}$$

So therefore $\frac{1}{4}\%$ is the same as the fraction $\frac{1}{400}$.

If you have a percentage like $12\frac{1}{2}\%$ you would change this fraction to an improper fraction first before you convert it.

So $12\frac{1}{2}\%$ is the same as $\frac{25}{2}\%$

Step 1. Change $\frac{25}{2}$ to a fraction over 100.

$$\frac{\frac{25}{2}}{100}$$

Step 2. Rewrite this as a division of fractions

$$= \frac{25}{2} \div \frac{100}{1}$$

Step 3. Change to the sign to multiplication and invert the second fraction

$$= \frac{25}{2} \times \frac{1}{100}$$

Step 4. Simplify

$$= \frac{\overset{1}{\cancel{25}}}{2} \times \frac{1}{\cancel{100}_{4}} = \frac{1}{8}$$

Therefore $12\frac{1}{2}\%$ is equal to $\frac{1}{8}$.

Convert these fractional percentages to fractions.

a) $\frac{2}{9}\%$

f) $2\frac{1}{4}\%$

b) $\frac{5}{6}\%$

g) $5\frac{2}{5}\%$

c) $\frac{1}{10}\%$

h) $11\frac{1}{2}\%$

d) $\frac{1}{5}\%$

i) $8\frac{3}{4}\%$

e) $\frac{2}{3}\%$

j) $7\frac{2}{7}\%$

Did you figure out the cheat for the above system? You just multiply the denominator by 100 and simplify. Make sure you change mixed numbers to improper fractions first.

Chapter 3

Percentages and Decimals

To **convert a percentage to a decimal**, just divide the amount by 100 or move the decimal point 2 places to the left and drop the % sign. Don't forget if there is no decimal point shown, you assume it is at the end of the last digit.

Examples,

a) Convert 5% to a decimal

Step 1. As there is no decimal point shown, you know it goes after the last digit. Therefore 5 becomes 5.0.

Step 2. Now just move the decimal point two places to the left and drop the % sign. Don't forget to add zeros to the left of the 5. (See Easy Steps Math Decimals).

So 5% becomes 0.05

b) Convert 342% to a decimal

Step 1. Place a decimal point after the last digit. Therefore 342 becomes 342.0

Step 2. Move the decimal point 2 places to the left and drop the % sign.

So 342% becomes 3.42

c) Convert 0.09% to a decimal.

Step 1. A decimal point already exists so there is no need to add one.

Step 2. Move the decimal point two places to the left and drop the % sign. Add extra zeros to the left of the number.

So 0.09% becomes 0.0009

You have a go at these questions.

Convert the following to percentages to decimals.

a) 23%

b) 78%

c) 163%

d) 247%

e) 879%

f) 1026%

g) 0.3%

h) 0.98%

i) 0.08%

j) 0.004%

If you want to **convert a decimal to a percentage**, you would do the opposite of what you did above, that is you would multiply by 100, or move the decimal point two places to the right. Remember to add the percentage sign.

Examples,

a) Convert 0.5 to a percentage

Step 1. Add extra zeros after the last digit (if it is necessary)

Step 2. Move the decimal point to the right two spaces and add $\%$.

So 0.5 becomes 50.0%

Since whole numbers don't need decimal points, you can write your answer as 50%

b) Convert 0.23 to a percentage.

Step 1. Add extra zeros after the last digit (if it is necessary)

Step 2. Move the decimal point to the right two spaces and add $\%$

So 0.23 becomes 23%

c) Convert 0.007 to a percentage

Step 1. Add extra zeros after the last digit (if it is necessary)

Step 2. Move the decimal point to the right two spaces and add %.

So 0.007 becomes 0.7%

Convert the following decimals to percentages.

a) 0.6

b) 0.1

c) 0.47

d) 0.01

e) 0.023

f) 0.07

g) 0.567

h) 5.6

i) 6

j) 2.02

Chapter 4

Showing One Amount as a Percentage of Another

There may be times when you need to work out **one amount as a percentage of another**. To do this you place the amount you want to turn into a percentage over the second amount and multiply by $\frac{100}{1}$.

Examples,

a) If there were 250 students enrolled in a school of which 130 are male what percentage of students are female?

b) If an athlete has completed 1200 meters of a 3 km run what percentage of the run has she completed?

The most important aspect of these questions is reading them carefully. The calculations are the easy part. Knowing what the question is asking is where many students go wrong. It is important to answer what the question is asking, not what you think the question is asking. Therefore read the whole question, more than once if necessary.

For a) above, the question is asking what percentage of female students are enrolled in the school, but the information given is about the number of male students. Your first step is to work out how many female students there are, and secondly you need to show this amount as a percentage of the total number of students.

So,

Step 1. Work out the number of female students. $250 - 130 = 120$ female students.

Step 2. Calculate the number of female students as a percentage of the total number of students. (Female students over total students times 100.)

$$\frac{120}{250} \times \frac{100}{1}$$

Step 3. Cross simplify

$$= \frac{120}{{}_{10}\cancel{250}} \times \frac{\cancel{100}^{4}}{1} = \frac{120}{10} \times \frac{4}{1}$$

Step 4. Multiply out the numerator then the denominator

$$= \frac{480}{10}$$

Step 5. Simplifying and add %.

$$= 48\%$$

Therefore the percentage of female students in the school is 48%

For question b) above, you will notice that the question has meters and kilometers mixed together. Before you do any calculations, you must make the units the same, so you must choose either meters or kilometers. In this instance meters is better. And you need to work out the distance the athlete has run as a percentage of the total distance.

Step 1. Make the units the same, i.e. 1200 m and 3000m

Step 2. Calculate the distance run as a percentage of the total distance. (Distance run over total distance times 100.)

$$\frac{1200}{3000} \times \frac{100}{1}$$

Step 3. Cross simplify,

$$= \frac{120\cancel{0}}{3\cancel{0}\cancel{00}} \times \frac{1\cancel{00}}{1} = \frac{120}{3} \times \frac{1}{1}$$ zeros are easy to cancel away from top and bottom, so long as you cancel the same number of zeros. In this case three from the top and three from the bottom.

Step 4. Multiply out the numerator then the denominator

$$=\frac{120}{3}$$

Step 5. Simplifying and add %.

$$=40\%$$

Therefore the percentage distance the athlete has run is 40%

For the following, show the first number as a percentage of the second number. Don't forget to make any necessary conversions first. Give answers to one decimal place where needed.

a) 15, 33

b) 21, 45

c) 450g, 2kg

d) 9mm, 3cm

e) 45 cents, $4.00

f) 90minutes, 3.5hours

g) 5kg, 600g

h) 4 days, 6 weeks

i) 375kg, 1tonne

j) 50cm, 400mm

Chapter 5

Complementary Percentages

Complementary Percentages are the easiest percentages to calculate. Complementary percentages add up to 100%.

For example, a question might ask: if 48% of students in a school are girls, then what percentage are boys? The answer is clearly 52%.

48% and 52% are complementary percentages because they add up to 100%.

Another question might ask: if fruit salad is made up of 23.6% oranges, 24.8% kiwi fruit, 29.5% apples, then what percentage is bananas? First add the figures you have then subtract this answer from 100. So 23.6 + 24.8 + 29.5 = 77.9%. Subtracting this from 100% you get 22.1%. So the answer is 22.1% of the fruit salad is bananas.

Worded questions should always have worded answers unless specified otherwise.

Find the complements of the following percentages.

a) 56%

b) 87%

c) 65%

d) 12%

e) 32.9%

f) 78.2%

g) 45.6 and 19.5

h) 18.3 and 72.4

i) 12.6 and 56.4 and 11.0

j) 32 and 45 and 8

Chapter 6

Finding the Percentage of an Amount

If you have a question like *Find 37% of 500*, you would use the information learned in the Easy Steps Math Fractions book together with the information learned in this book to come up with a solution.

Step 1. Rewrite the expression above in mathematical terms.

37% is the same as $\frac{37}{100}$; '*of*' is the same as times $(\times)$ and 500 is the same as $\frac{500}{1}$.

Putting it all together you get

$$\frac{37}{100} \times \frac{500}{1}$$

Step 2. Cross simplify

$$= \frac{37}{\cancel{100}} \times \frac{5\cancel{00}}{1} = \frac{37}{1} \times \frac{5}{1}$$

Step 3. Multiply the fractions

$$\frac{185}{1}$$

Step 4. Simplify

$$185$$

Therefore 37% of 500 is 185

Here is another example. *Find* $45\frac{1}{2}\%$ *of 300*

Step 1. Rewrite mathematically

$$\frac{91}{200} \times \frac{300}{1}$$

Step 2. Cross simplify

$$\frac{91}{2\cancel{00}} \times \frac{3\cancel{00}}{1} = \frac{91}{2} \times \frac{3}{1}$$

Step 3. Multiply the fractions

$$\frac{273}{2}$$

Step 4. Simplify

$$136\frac{1}{2}$$

So $45\frac{1}{2}\%$ *of 300* is $136\frac{1}{2}$.

Try these questions

a) 80% of 16

b) 25% of 35

c) 17% of 65

d) 230% of 62

e) 147% of 123

f) $15\frac{1}{2}\%$ of 800

g) $5\frac{1}{4}\%$ of 9000

h) $1\frac{5}{9}\%$ of 1630

i) $4\frac{5}{6}\%$ of 650

j) $3\frac{1}{7}\%$ of 2200

Sometimes finding the **percentage of an amount using decimals** is an easier process for calculations.

For example, *find 25.6% of 250*

Remember that to convert a percentage to a decimal, you need to divide the percentage by 100 (or move the decimal point to the left 2 places).

Therefore to find 25.6% of 250

Step 1. Rewrite the question using decimals

0.256×250

Step 2. Multiply

= 64

So 25.6% of 250 is 64

You try these questions. Convert the percentage to a decimal first and give your answers to 2 decimal places.

a) 80% of 12

b) 65% of 90

c) 20% of 35

d) 2.3% of 15

e) 3.7% of 43

f) 0.3% of 10

g) 5.2% of 120

h) 66% of 480

i) 25% of $44.60

j) 92% of $1235

Chapter 7

Application of Percentages

There are many areas of life where you will need percentages. When you go shopping, various stores offer discounts. If you own a business you would need to mark up the price of the items you have bought from the supplier in order to work out a selling price. Also as a business owner you may want to calculate your percentage profit or maybe percentage loss if you're not making a profit yet. Banks charge interest when you borrow money from them, and give you interest when you deposit money with them. Interest is calculated as a percentage of money owing or deposited. Some sales people earn their income through commissions and bonuses, and this is usually a percentage of what they sell. So as you can see, percentages are everywhere and it's important to know how to use them.

Examples

a) You've gone shopping with your friends and you see an outfit you've had your eye on for a few weeks and it has been discounted by 25%. Its original price was $125.00. How much are you going to pay? You can answer this in two ways.

Firstly you know the discount is 25% therefore you will be paying the complement of 25%, which is 75%. So just work out 75% of $125.00 which is $93.75.

$100\% - 25\% = 75\%$

$75\% \times \$125.00 = \$93.75$

The other way is to work it out is to calculate 25% of $125.00, which is $31.25, and then subtract this from $125.00. This gives you $93.75.

$25\% \times \$125.00 = \$31.25$

$\$125.00 - \$31.25 = \$93.75$

Either way, the answer is the same.

b) If you are a shop owner and you sell stereo equipment and you bought a CD player from your supplier for $110.00 and it has a 30% markup, how much will you sell it for?

First calculate 30% of $110, and then add this answer to the $110.

$30\% \times \$110.00 = \$33.00$

$\$33.00 + \$110.00 = \$143.00$

So the selling price or retail price of the CD player is $143.00.

c) To calculate profit, you must subtract cost price from selling price. So the rule is Profit =Selling Price - Cost Price. If this figure is a positive number, that is a number larger then zero, then you have made a profit, if the figure is a negative number, then you have made a loss. But this only gives you a dollar figure and not a percentage. To calculate your percentage profit or loss you need the formula $\%Profit = \frac{Profit}{Cost} \times \frac{100}{1}$ or the formula

$\%Loss = \frac{Loss}{Cost} \times \frac{100}{1}$.

Using the CD player example above, if you sold the player for its retail price, then, using the rule Profit = Selling Price – Cost Price you would have:

$143.00 – $110.00 = $33.00 (the mark up amount).

To calculate the percentage profit, use the formula:

$$\%Profit = \frac{Profit}{Cost} \times \frac{100}{1}$$

$$\%Profit = \frac{33}{11\cancel{0}} \times \frac{10\cancel{0}}{1}$$

$$\%Profit = \frac{330}{11}$$

$$\%Profit = 30\%$$

If you sold the player for $90, then your figures would look like this:

Profit = Selling Price – Cost Price

$90.00 – $110.00 = -$20.00

So you have a loss of $20.00 or a profit of -$20.00

To calculate the percentage loss, use the formula:

$$\%\text{Loss} = \frac{\text{Loss}}{\text{Cost}} \times \frac{100}{1}$$

$$\%\text{Loss} = \frac{20}{110} \times \frac{100}{1}$$

$$\%\text{Loss} = \frac{2\cancel{0}}{11\cancel{0}} \times \frac{100}{1}$$

$$\%\text{Loss} = \frac{200}{11}$$

$$\%\text{Loss} = 18.2\%$$

d) If you have $10,000.00 saved in a bank, which is earning you interest at a rate of 4% per annum (per year). Then to work out how much money you would have after one year, you would work out the amount you would receive in interest then add it to the money you put in the account. For instance:

4% × $10,000 = $400

$10,000 + $400 = $10,400

So after one year you would have an extra $400 in your account.

If you borrowed $300,000 to buy a house at an interest rate of 4% per annum, then you would work out how much interest you would have to pay back the bank. For instance

4% x $300,000 = $12,000

So you would be paying back $12,000 per year to the bank. Of course you would do this monthly to make it easier so $1000 per month.

e) Some people's jobs earn them a commission for their income. These are usually sales jobs and the commission is a percentage of what they sell. If you are a real estate agent and you sold a house for a client, the agency may receive a commission of 2-3%. So if a house sold for $500,000 and the agent's commission is 2%, then the seller will have to pay the real estate agent $10,000 for having sold the property.

2% x $500,000 = $10,000.

Multiplication Tables

To make calculations really easy, learn your multiplications tables. Here is a set of multiplication tables from 1 x 1 to 12 x 12 to help you if you need it.

1 x 1 = 1	2 x 1 = 2	3 x 1 = 3	4 x 1 = 4
1 x 2 = 2	2 x 2 = 4	3 x 2 = 6	4 x 2 = 8
1 x 3 = 3	2 x 3 = 6	3 x 3 = 9	4 x 3 = 12
1 x 4 = 4	2 x 4 = 8	3 x 4 = 12	4 x 4 = 16
1 x 5 = 5	2 x 5 = 10	3 x 5 = 15	4 x 5 = 20
1 x 6 = 6	2 x 6 = 12	3 x 6 = 18	4 x 6 = 24
1 x 7 = 7	2 x 7 = 14	3 x 7 = 21	4 x 7 = 28
1 x 8 = 8	2 x 8 = 16	3 x 8 = 24	4 x 8 = 32
1 x 9 = 9	2 x 9 = 18	3 x 9 = 27	4 x 9 = 36
1 x 10 = 10	2 x 10 = 20	3 x 10 = 30	4 x 10 = 40
1 x 11 = 11	2 x 11 = 22	3 x 11 = 33	4 x 11 = 44
1 x 12 = 12	2 x 12 = 24	3 x 12 = 36	4 x 12 = 48

5 x 1 = 5	6 x 1 = 6	7 x 1 = 7	8 x 1 = 8
5 x 2 = 10	6 x 2 = 12	7 x 2 = 14	8 x 2 = 16
5 x 3 = 15	6 x 3 = 18	7 x 3 = 21	8 x 3 = 24
5 x 4 = 20	6 x 4 = 24	7 x 4 = 28	8 x 4 = 32
5 x 5 = 25	6 x 5 = 30	7 x 5 = 35	8 x 5 = 40
5 x 6 = 30	6 x 6 = 36	7 x 6 = 42	8 x 6 = 48
5 x 7 = 35	6 x 7 = 42	7 x 7 = 49	8 x 7 = 56
5 x 8 = 40	6 x 8 = 48	7 x 8 = 56	8 x 8 = 64
5 x 9 = 45	6 x 9 = 54	7 x 9 = 63	8 x 9 = 72
5 x 10 = 50	6 x 10 = 60	7 x 10 = 70	8 x 10 = 80
5 x 11 = 55	6 x 11 = 66	7 x 11 = 77	8 x 11 = 88
5 x 12 = 60	6 x 12 = 72	7 x 12 = 84	8 x 12 = 96

9 x 1 = 9	10 x 1 = 10	11 x 1 = 11	12 x 1 = 12
9 x 2 = 18	10 x 2 = 20	11 x 2 = 22	12 x 2 = 24
9 x 3 = 27	10 x 3 = 30	11 x 3 = 33	12 x 3 = 36
9 x 4 = 35	10 x 4 = 40	11 x 4 = 44	12 x 4 = 48
9 x 5 = 45	10 x 5 = 50	11 x 5 = 55	12 x 5 = 60
9 x 6 = 54	10 x 6 = 60	11 x 6 = 66	12 x 6 = 72
9 x 7 = 63	10 x 7 = 70	11 x 7 = 77	12 x 7 = 84
9 x 8 = 72	10 x 8 = 80	11 x 8 = 88	12 x 8 = 96
9 x 9 = 81	10 x 9 = 90	11 x 9 = 99	12 x 9 = 108
9 x 10 = 90	10 x 10 = 100	11 x 10 =110	12 x 10 = 120
9 x 11 = 99	10 x 11 = 110	11 x 11 = 121	12 x 11 = 132
9 x 12 = 108	10 x 12 = 120	11 x 12 = 132	12 x 12 = 144

Answers

Converting a Percentage to a Fraction

a) $\frac{11}{100}$ b) $\frac{67}{100}$ c) $\frac{23}{100}$ d) $\frac{9}{25}$ e) $1\frac{7}{10}$ f) $1\frac{29}{100}$
g) $3\frac{3}{20}$ h) $2\frac{2}{5}$ i) $3\frac{3}{5}$ j) 4

Converting a Fraction to a Percentage

a) 90% b) 35% c) 50% d) 48% e) 30% f) 70% g) 175%
h) 6% i) 80% j) 125%

Converting a Fractional Percentage to a Fraction

a) $\frac{1}{450}$ b) $\frac{1}{120}$ c) $\frac{1}{1000}$ d) $\frac{1}{500}$ e) $\frac{1}{150}$ f)
$\frac{9}{400}$ g) $\frac{27}{500}$ h) $\frac{23}{200}$ i) $\frac{7}{80}$ j) $\frac{51}{700}$

Converting a Percentage to a Decimal

a) 0.23 b) 0.78 c) 1.63 d) 2.47 e) 8.79 f) 10.26 g) 0.003
h) 0.0098 i) 0.0008 j) 0.00004

Converting a Decimal to a Percentage

a) 60% b) 10% c) 47% d) 1% e) 2.3% f) 7% g) 56.7% h) 560% i) 600% j) 202%

Showing One Amount as a Percentage of Another Amount

a) 45.5% b) 46.7% c) 22.5% d) 30% e) 11.3% f) 42.9% g) 833.3% h) 9.5% i) 37.5% j) 125%

Complementary Percentages

a) 44% b) 13% c) 35% d) 88% e) 67.1% f) 21.8% g) 34.9% h) 9.3% i) 20% j) 15%

Finding the Percentage of an Amount

a) $12\frac{4}{5}$ b) $8\frac{3}{4}$ c) $11\frac{1}{20}$ d) $142\frac{3}{5}$ e) $180\frac{81}{100}$

f) 124 g) $472\frac{1}{2}$ h) $25\frac{16}{45}$ i) $31\frac{5}{12}$ j) $69\frac{1}{7}$

Finding the Percentage of an Amount Using Decimals

a) 9.60 b) 58.50 c) 7.00 d) 0.35 e) 1.59 f) 0.03 g) 6.24 h) 316.80 i) $11.15 j) $1136.20

Glossary of Useful Terms

Canceling zeros: This is the process where zeros in the numerator and denominator can be cancelled. For each zero in the numerator a zero in the denominator can be cancelled. This can occur in a single fraction or in fractions being multiplied.

For example.

For the single fraction $\frac{120}{200}$ the zero in the numerator can cancel with one zero in the denominator.

$$\frac{12\not{0}}{20\not{0}}$$

This is a shortcut for dividing the top and bottom of the fraction by 10. The equivalent fraction then becomes $\frac{12}{20}$ which of course simplifies to $\frac{3}{5}$.

For fractions being multiplied such as $\frac{35}{140} \times \frac{60}{75}$ you can cancel the zero in the 60 with the zero in the 140 $\frac{35}{14\not{0}} \times \frac{6\not{0}}{75}$ and then continue to simplify any other numbers.

Sum refers to addition. The sum of two numbers is the answer of one number **plus** another number. E.g. the sum of 2 and 6 is 8, $(2 + 6 = 8)$.

Difference refers to subtraction. The difference between two numbers is the answer of one number **minus** another number. E.g. the difference between 6 and 2 is 4, $(6 - 2 = 4)$.

Product refers to multiplication. The product of two numbers is the answer of one number **times** another number. E.g. the product of 2 and 6 is 12, $(2 \times 6 = 12)$.

Quotient refers to division. The quotient is the answer of one number being **divided** by another number. E.g. the quotient of 6 and 2 is 3 $(6 \div 2 = 3)$.

Thank you for reading!

Dear Reader,

I hope you found this **Easy Steps Math – Percentages** book useful, either for yourself or for your children.

The **Easy Steps Math** series began as a set of math notes that I used in class for my students to copy from the board. I found that doing this helped the students in at least two ways. Firstly, with their homework, because they didn't forget how to do the work that was explained in class, and secondly, with their results, because they used the notes to study for tests and exams.

Where students in other classes were getting detentions for non-completion of homework, my students were getting homework done, their results were improving and they were enjoying math.

Students from other classes, even older students, were thanking me for my notes, as they were copying them from their peers because they found them so easy to follow and learn from. During a parent/teacher conference, one parent also thanked me because of how his child was able to easily learn the work, and that he, as a teacher, was using my notes in his classes, in his school.

This is when I realised that these notes would benefit many more students if they were published. Thus we are at this point.

I welcome any comments you have about this **Percentages** book. Tell me what you liked, loved, or even hated about it. I'd be happy to hear from you. You can email me at robwatchman@gmail.com

Finally, I would like to ask a favour. I would appreciate it if you would write a review of this book so that others can get an idea of how helpful it may be for them or their children. You would be aware that reviews are hard to come by because many readers don't go back to where they purchased their books.

So if you have the time, I would appreciate a few comments.

Thank you so much for reading the **Easy Steps Math – Percentages** book and for spending your time with me.

In Gratitude,

Robert Watchman

Made in the USA
Las Vegas, NV
01 November 2024